Space Junk

Story by Heather Hammonds
Illustrations by Millie Liu

Contents

Chapter 1

A Light in the Sky

Aiden was spending a week with his cousin Krista and his Aunt Kelly and Uncle Ralph. They lived on a farm in the country.

Aiden loved visiting the farm.

In the evenings, Aiden liked to go outside and look up at the starry sky. He wanted to be an astronomer when he was older.

"I can see lots of stars, here," he told Krista. "At home in the city, all the lights make the stars hard to see."

One evening, after dinner,
Aiden and Krista were out in the garden.
They were looking at the full moon,
which was beginning to rise.

“It’s so cold!” said Krista. “I’m going inside to get my coat.”

Aiden stayed outside.
He watched the big, yellow moon rise high over the farm.

Suddenly, Aiden saw a bright object race across the sky. Then he heard a loud **BANG**!

"Krista, Uncle Ralph, Aunt Kelly!" he yelled. "I think something from space might have crashed on the farm!"

Everyone ran outside.

“It sounded like it landed in a field over there,” Aiden told them, pointing.

“I can hear the cows,” said Uncle Ralph. “Something has disturbed them. Let’s go and take a look.”

Chapter 2

Night Hunt

Everyone got into the truck
and Uncle Ralph drove across the field.

"Maybe it's a meteorite," said Aunt Kelly.

"What's a meteorite?" asked Krista.

"It's a bit of rock that has fallen to Earth from space," Aiden replied, excitedly.

Aiden looked through the truck's window.
The headlights shone on the cows,
but he couldn't see the thing that had fallen from the sky.

Suddenly, Aiden pointed to a big shed
in the middle of the field.
"There's a hole in the roof!" he said.

Uncle Ralph stopped the truck.
Everyone jumped out and followed Aunt Kelly
as she ran over to the shed.

Aunt Kelly put her phone flashlight on
and shone it into the shed.

Aiden pointed to a small piece of metal on the ground.
"Look, Aunt Kelly!" he said.
"That must be the thing that fell from space.
Can you take a photo?"

Aunt Kelly nodded.

“I hope that thing is not dangerous,” said Uncle Ralph. “Don’t get too close.”

“Perhaps it fell off a plane,” said Aunt Kelly. “I’m going to call the police.”

Chapter 3

The Mysterious Object

Everyone walked back to the truck and waited for the police.

"I don't think that bit of metal fell off a plane," whispered Aiden.
"It flew through the air so fast!"

"Maybe the metal was made by aliens," said Krista.

"That only happens in movies," Aiden replied.

Soon, a police officer arrived.
Aiden showed her the piece of metal.

"We'll ask some scientists from the Space Agency to come and look at this tomorrow," she said. "Please don't touch it."

That night, Aiden lay in bed
and thought about meeting the scientists
from the Space Agency.

He hoped they could tell him
exactly what the mysterious object was.

Chapter 4

Finding Out

The next morning, three scientists from the Space Agency arrived at the big shed.

"Who saw the object fall from the sky?" asked Carla, one of the scientists.

"I did," said Aiden. "I heard a big bang, too."

"Do you know what it is?" asked Aiden.

"It's a piece of space junk," replied Carla.

"What's space junk?" asked Krista.

"Space junk is mostly made up of old bits of machines that were sent into space," said Carla.
"It looks like this piece of junk has come from a satellite."

Aiden and Krista were amazed.

"Sometimes, space junk falls back to Earth," said Carla.
"It gets very hot on its way down, and usually it burns up.
Bigger bits like this one can hit the ground."

Just then, a TV news van arrived.

"I saw your mum send the photo of the space junk to her friends," Aiden whispered to Krista. "One of them must have contacted the TV station."

A reporter came over and asked Aiden and Krista all about what had happened.
They told him about the space junk and how they had found it.

Then Carla told the reporter that as more machines are sent into space, more rubbish is left floating around Earth. "Scientists are working on ways to help clean up space junk," she said.

Aiden asked Carla if he could keep the piece of space junk, but she told him it belonged to the country that had sent it into space.

"We'll try to find out who owns it and give it back to them," she said.

That night, Aiden and Krista were on the TV news.

"We're famous!" said Krista.

"I still wish we could have kept the space junk," said Aiden. "But I'm so glad it fell to Earth right here on the farm."